Dogs shake.

# All Clean!

by Harriet Ziefert
pictures by Henrik Drescher

Harper & Row, Publishers

Cats wash.

Elephants

shower.

Hippos soak.

Birds splash.

Monkeys pick.

Lions lick.

Rhinos . . .

rub.

Gorillas scratch.

Pandas scrub.

Flamingos preen.

Snakes shed.

Beavers bathe.

# Fish are always clean!